ORGANIZATION OF DNA
IN CHROMATIN

Rather than bending uniformly along its length, nucleosomal DNA is proposed to consist of multiple segments of B- and A- DNA held together by kinks when forming its left-handed toroidal superhelical structure

Henry M. Sobell

ISBN: 979-8-88615-112-1 (Paperback)
            979-8-88615-113-8 (Hardback)

Front Cover Illustration:

The 'mixed-puckered' kinked structure that gives rise to the left-handed toroidal superhelix in nucleosomal DNA. Central red line indicates the superhelix axis; blue lines indicate helical-axes present within A- and B- DNA segments connected together by kinks in this left-handed toroidal superhelical structure.

INKS &
BINDINGS

Inks and Bindings
888-290-5218
www.inksandbindings.com
orders@inksandbindings.com

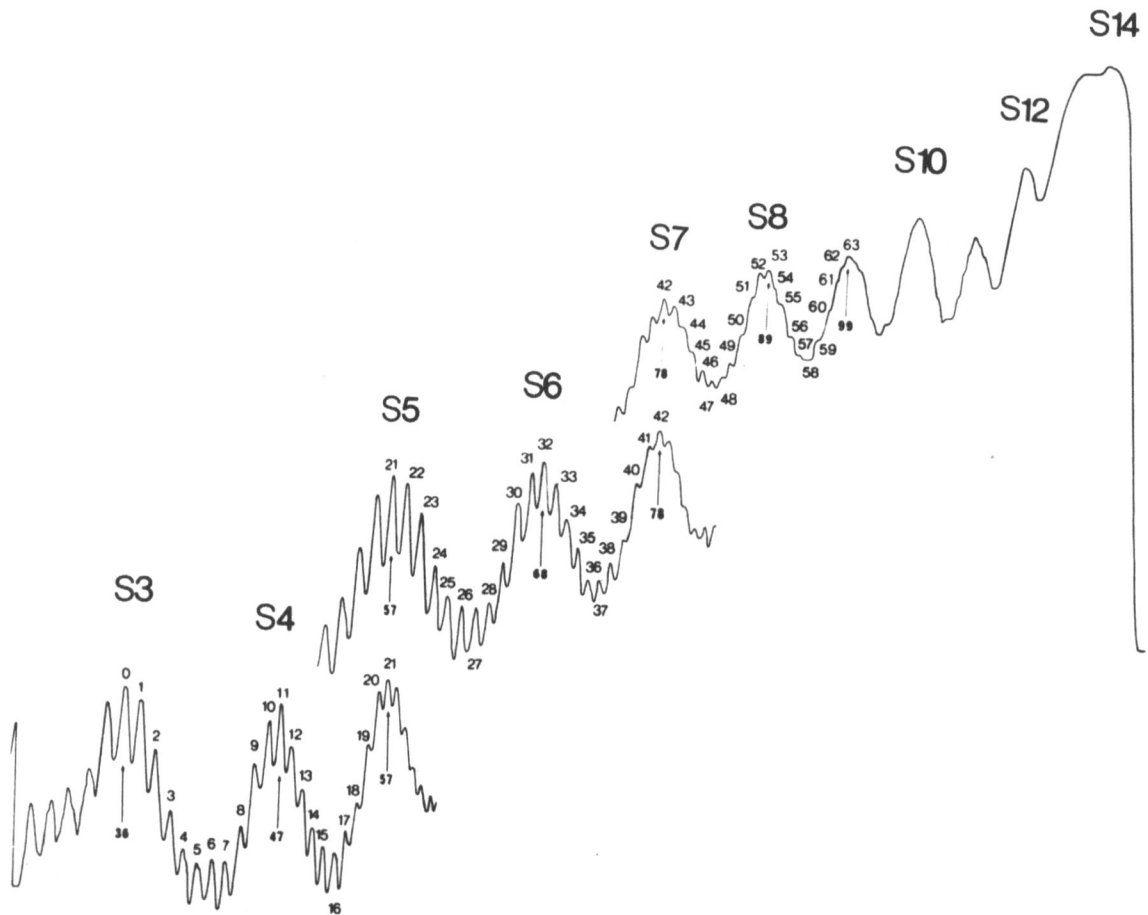

This densitometer tracing shows an electrophoretic pattern of oligonucleotides arising after pancreatic DNase I digestion of a collection of DNA molecules with random sequences, each containing 147 base-pairs and labelled at their 5' ends with radioactive phosphorous -- in the presence of calcium phosphate crystals. The tracing shows a series of maxima spaced 10.5 base-pairs apart; these, in addition, containing finer peaks that differ by one nucleotide. Similar patterns have been obtained with nucleosomal DNA when it is complexed to histone-octamer cores in nucleosomes.

Rather than bending uniformly along its length, the author proposes these observations to reflect the presence of a left-handed toroidal superhelical structure being composed of multiple segments, each containing 10 base-pairs of B- DNA or 11 base-pairs of A- DNA, these being held together by "mixed-puckered kinks". In both cases, probability considerations predict cutting patterns to be symmetrically distributed around integral multiples of 10.5 base-pairs along DNA, the relative magnitudes of the surrounding peaks in these patterns being governed by the binomial distribution.

Preface

This paper demonstrates that -- provided there were an equal probability that either 10 base-pairs of B-DNA or 11 base-pairs of A-DNA exist within any given segment of the left-handed toroidal superhelix in nucleosomal DNA, these being connected together by "mixed- puckered" kinks in its structure -- then a population of such *aperiodic* structures is expected to give rise to the *periodic* cutting-patterns observed experimentally. This would be true for naked DNA molecules immobilized on a calcium-phosphate crystalline surface as well, provided they also form left -handed toroidal superhelices under such conditions. In both cases, probability considerations predict cutting patterns to be symmetrically distributed around integral multiples of 10.5 base-pairs along DNA -- the relative magnitudes of the surrounding peaks in these patterns being governed by the *binomial-distribution*. We describe the possible implications this model has in understanding the higher-order structural organization of DNA in chromatin.

HENRY M. SOBELL
Lake Luzerne, New York
January, 2023

Table of Contents

1. Introduction

Although it is now widely accepted that nucleosomal DNA forms a left-handed toroidal superhelix when winding around the nucleosome core particle, little is known about the flexibility in its structure that allows this to happen.

Earlier, I put forward a kinked model to understand how DNA is organized within the nucleosome (1, 2-5). The model assumed nucleosomal DNA to be in its B-form, separated by 'mixed- puckered kinks' every 10 base-pairs. Here, I present a modification to this model, this being necessary to explain important additional experimental information uncovered several years after the model was proposed (6-10).

The modified model proposes that if there were an equal probability that both 10 base-pairs of B-DNA or 11 base-pairs of A-DNA exist within any given segment of the left-handed toroidal superhelical structure -- these being connected together by mixed-puckered kinks -- then a population of such aperiodic structures is expected to give rise to the periodic cutting-patterns observed experimentally. This would be true for naked DNA molecules immobilized on a calcium-phosphate crystalline surface as well, provided they also formed left-handed toroidal superhelices under these conditions. In both cases, probability considerations predict cutting patterns to be symmetrically distributed around integral multiples of 10.5 base-pairs along DNA, the relative magnitudes of the surrounding peaks in these patterns being governed by the binomial distribution.

An important prediction made by this revised model concerns the number of base-pairs present within any given nucleosome. The model predicts this number to vary (i.e., lying between 140 and 154 base-pairs; however, having the highest probability that it contains 147 base-pairs), the magnitudes of the surrounding peaks being governed by the binomial distribution.

2. Presence of Anisotropic Flexibility in DNA Structure

We begin by reviewing the properties of space-filling CPK models of DNA, which reveal the presence of a highly directional (i.e., anisotropic) flexibility in its structure. This anisotropy becomes evident upon examining these models in the absence of its aluminum helix- axis, whose presence imparts an artificial rigidity to its structure.

See Figure 1.

If one begins by bending DNA towards its wide groove direction (i.e., down the 2-fold symmetry axis lying between adjacent base-pairs, this being perpendicular to its helix-axis), base-pairs begin to "roll" on each others van der Waals surfaces, this being accompanied by small but systematic alterations in the torsional angles defining the sugar-puckering, base-sugar orientation and sugar-phosphate linkages in DNA.These alterations allow DNA to bend into its wide groove, eventually resulting in the formation of the mixed-puckered kink [i.e., C3' endo (3' - 5') C2' endo], visible from the narrow groove.

CPK MODELS ILLUSTRATING THE ANISOTROPIC FLEXIBILITY OF DNA

H.M. Sobell, C. -C. Tsai, S.C. Jain and S.G. Gilbert, J. Mol. Biol. 114, 333-365 (1977)

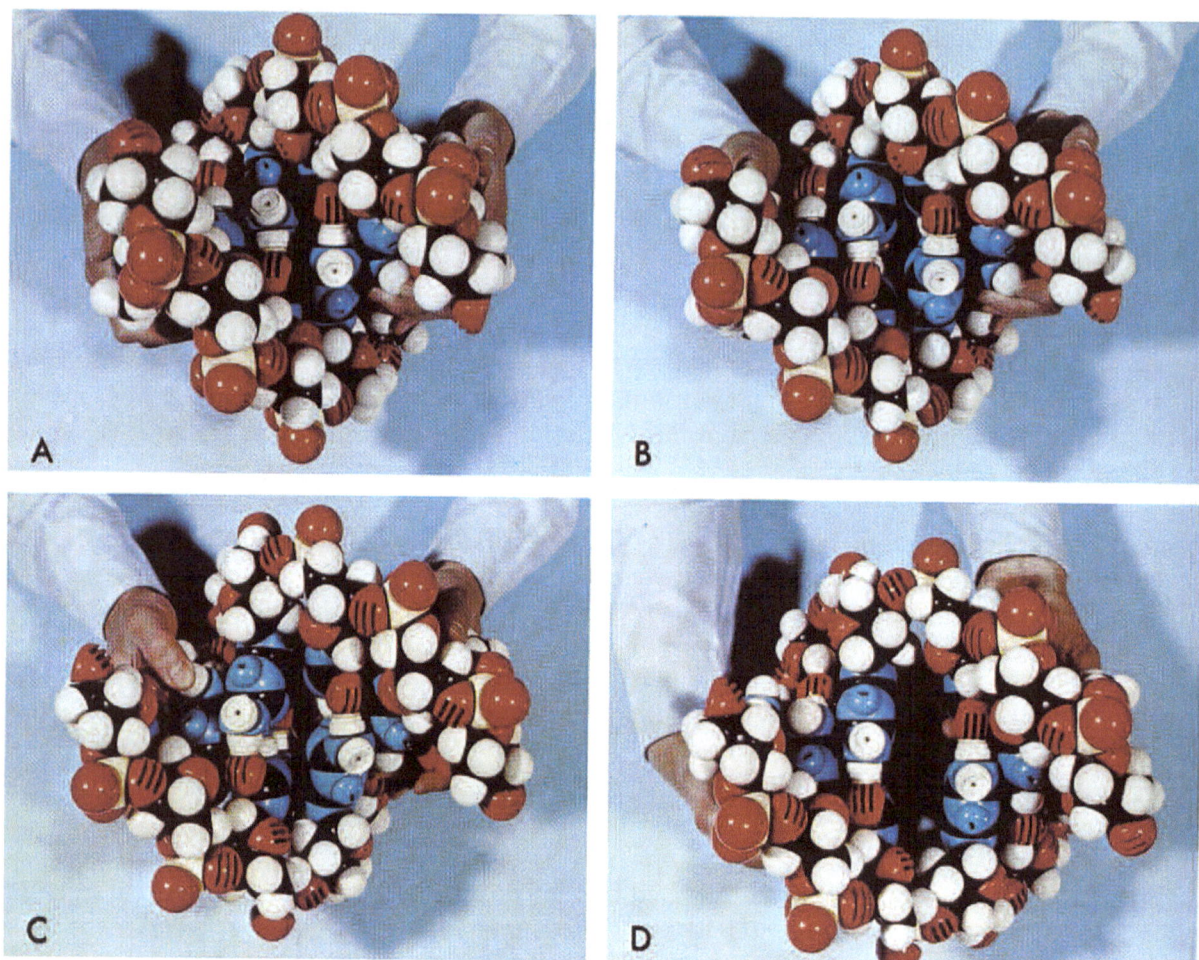

Figure 1: Space-filling Corey-Pauling-Koltun (CPK) molecular models of DNA used to demonstrate the anisotropic flexibility that is present down its 2-fold axis of symmetry that passes between adjacent base-pairs.

(A)   B- DNA viewed from the narrow groove direction down its 2-fold axis, passing between adjacent base-pairs.

(B)   B- DNA flexed into its major-groove direction, resulting in a "roll angle" of about 15 degrees between base-pairs, accompanied by a flattening of the sugar-pucker on the 5' side of the base-paired dinucleotide-structure, combined with a small decrease in its chi torsional angle.

(C)   The appearance of the 'mixed-puckered kink', characterized by the formation of a C3' endo (3 - 5') C2' endo mixed sugar-puckering pattern -- surrounded by B- DNA on either site. Adjacent base-pairs are now partially-unstacked, forming a "roll-angle" of about 40 degrees and unwound -12 degrees, relative to B-DNA.

(D)   The flexible-hinge that results, allowing B- DNA to straighten and adjacent base-pairs to separate an additional 3.4 Angstroms. This localized conformational change is accompanied by additional unwinding, the total now being -26 degrees.

ORGANIZATION OF DNA IN CHROMATIN

H.M. Sobell, C.C- Tsai, S.G. Gilbert, S.C. Jain and T.D. Sakore, Proc. Natl. Acad. Sci. USA 73, 3068-3072 (1976)

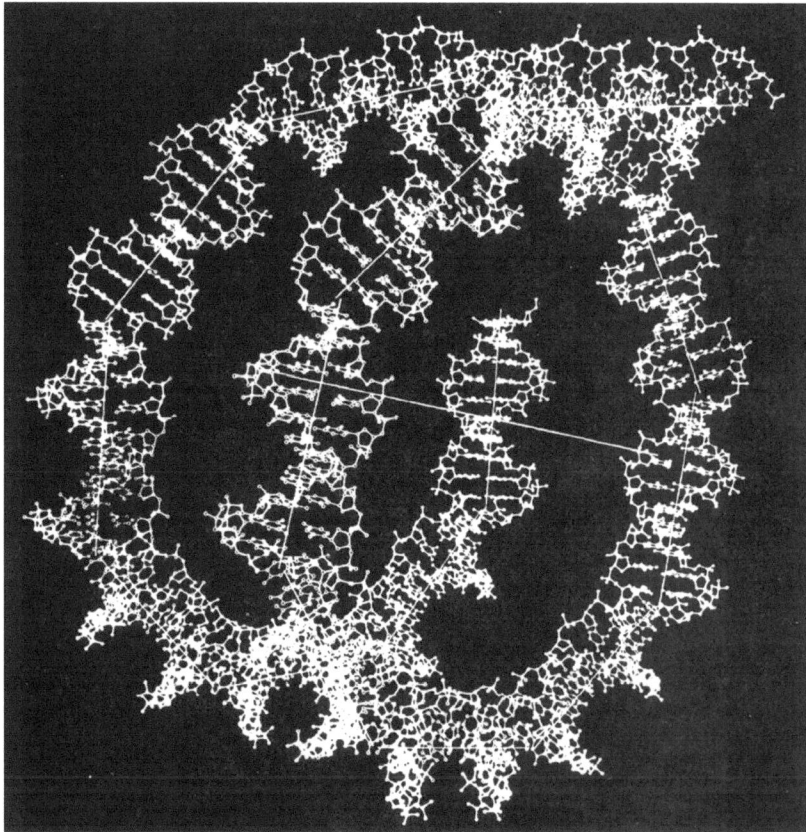

Figure 2: The original model put forward to understand the organization of DNA within the nucleosome -- the 'mixed-puckered kink' appears every 10-base pairs in B-DNA as it winds around the nucleosome histone-core to relieve the bending strain-energy that would otherwise result while forming this left-handed toroidal super-helical structure. The superhelix is generated from this ten base-pair containing asymmetric-unit, by a twist of -41.1 degrees and translation along the superhelix-axis of 5.26 Angstroms. The structure has a diameter of 100 Angstroms, and contains about one and one-half turns per 140 base-pairs. The long central-line indicates the superhelix-axis -- the length shown is 90 Angstroms. The information in this figure should be correlated with the information in the previous figure, Figure 1.

If, however there were an equal probability that either 11 base-pairs of A-DNA or 10 base-pairs of B-DNA exist within any given segment of the left-handed toroidal superhelical structure shown above, a population of such *aperiodic* structures can give rise to the *periodic* cutting-patterns observed experimentally (it is interesting to note in this regard that an A-DNA allomorph is known that contains 11 base-pairs per turn in 31.0 Angstroms (12)). This would be true for naked DNA molecules immobilized on a calcium-phosphate crystalline-surface as well, provided they also form left-handed toroidal super-helices under such conditions. In both cases, probability-considerations predict cutting patterns to be symmetrically distributed around integral multiples of 10.5 base-pairs along DNA, the relative magnitudes of the surrounding peaks in these patterns being governed by the binomial-distribution. I now propose this modified model to understand how DNA is organized within the nucleosome.

A

$(a + b)^1 = a + b$ [1 1]

$(a + b)^2 = a^2 + 2ab + b^2$ [1 2 1]

$(a + b)^3 = a^3 + 3a^2b + 3ab^2 + b^3$ [1 3 3 1]

$(a + b)^4 = a^4 + 4a^3b + 6a^2b^2 + 4ab^3 + b^4$ [1 4 6 4 1]

$(a + b)^5 = a^5 + 5a^4b + 10a^3b^2 + 10a^2b^3 + 5ab^4 + b^5$ [1 5 10 10 5 1]

$(a + b)^6 = a^6 + 6a^5b + 15a^4b^2 + 20a^3b^3 + 15a^2b^4 + 6ab^5 + b^6$ [1 6 15 201561]

B

```
                    1   1
                  1   2   1
                1   3   3   1
              1   4   6   4   1
            1   5   10   10   5   1
          1   6   15   20   15   6   1
      -----------------------------------------
          1   7   21   35   35   21   7   1
        1   8   28   56   70   56   28   8   1
      1   9   36   84   126   126   84   36   9   1
    1   10   45   120   210   252   210   120   45   10   1
   1   11   55   165   330   462   462   330   165   55   11   1
  1  12  66   220  495  792  924  792  495  220   66  12  1
 1  13  78  286  715  1287  1716  1716  1287  715  286  78  13  1
1  14  91  364  1001  2002  3003  3432  3003  2002  1001  364  91  14  1
```

2^n = 2 4 8 16 32 64 128 256 512 1,024 2,048 4,096 8,192 16,384

n = 1 2 3 4 5 6 7 8 9 10 11 12 13 14

Table 1: A. Algebraic explanation of the binomial theorem, and the origin of the binomial coefficients. B. Pascal's triangle, indicating how these binomial coefficients can be extended indefinitely, without having to carry out further algebra.

The presence of this mixed-puckered kink connecting segments containing 10 base-pairs of B- DNA or 11 base-pairs of A- DNA causes nucleosomal DNA to bend and to unwind, giving rise to a left-handed toroidal superhelical structure very similar to that originally proposed (i.e., an A DNA allomorph is known that contains 11 base-pairs per turn in 31.0 Angstroms (12)).

See Figure 2.

I will next explain how this modified model explains the 10.5 base-pair periodicity observed in electrophoretic patterns of nucleosomal DNA after digestion with the pancreatic DNase I. We first begin with a brief review of the binomial theorem.

3. A Brief Review of the Binomial Theorem

The binomial expression (a + b) raised to any power n (where n = 1, 2, 3, ...) leads to a series of polynomials, each having numerical coefficients defined as the binomial coefficients (see Table 1A).

These coefficients can be arranged in the form of a triangle (known as Pascal's triangle), which permit their values to be readily extrapolated to any value of n, without having to carry out further algebra (see Table 1B).

Thus, for example:

(n = 1)   [1   1] leads to (n = 2) [1   2   1], since 1+1 =2

(n = 2)   [1   2   1] leads to (n = 3) [1   3   3   1] , since 1+2=3   2+1=3

(n = 3)   [1   3   3   1] leads to (n = 4) [1   4   6   4   1] , since 1+3 =4

3+3=6 3+1=4

(n = 4)   [1   4   6   4   1] leads to (n = 5) [1   5   10   10   5   1] , since

1+4=5 4+6=10 6+4=10 4+1=5

and so on.

Once Pascal's triangle has been calculated, the binomial coefficients can be normalized as follows:

Thus, for example:

(n = 1)   [1   1] 2^1=2   (1/2 1/2) , or [0.5000   0.5000]

(n = 2)   [1   2   1] 2^2 = 4 (1/4 2/4 1/4) , or [0.2500   0.5000 0.2500]

(n − 3)   [1   3   3   1] 2^3=8 (1/8   3/8   3/8 1/8), or [0.1250 0.3750 0.3750 0.1250]

(n = 4)   [1   4   6   4   1] 2^4 = 16   (1/16 4/16 6/16 4/16 1/16), or [0.0625 0.2500

0.3750 0.3750 0.2500 0.0625]

and so on.

These normalized values of the binomial coefficients have been tabulated in Table 2, their sums being 1.0000 for each value of n. For reasons we next discuss, they will be used in the analysis which follows.

0.5000 0.5000

0.2500 0.5000 0.2500

0.1250 0.3750 0.3750 0.1250

0.0625 0.2500 0.3750 0.2500 0.0625

0.0312 0.1562 0.3125 0.3125 0.1562 0.0312

0.0156 0.0936 0.2343 0.3125 0.2343 0.0936 0.0156

0.0078 0.0546 0.1640 0.2734 0.2734 0.1640 0.0546 0.0078

0.0039 0.0312 0.1093 0.2187 0.2734 0.2187 0.1093 0.0312 0.0039

0.0019 0.0175 0.0703 0.1640 0.2460 0.2460 0.1640 0.0703 0.0175 0.0019

0.0009 0.009 0.0439 0.1171 0.2050 0.2460 0.2050 0.1171 0.0439 0.0097 0.0009

0.0004 0.0053 0.0268 0.0805 0.1611 0.2255 0.2255 0.1611 0.0805 0.0268 0.0053 0.0004

0.0002 0.0029 0.0161 0.0537 0.1208 0.1933 0.2255 0.1933 0.1208 0.0537 0.0161 0.0029 0.0002

0.0001 0.0015 0.0095 0.0349 0.0872 0.1571 0.2094 0.2094 0.1571 0.0872 0.0349 0.0095 0.0015 0.0001

0.0000 0.0008 0.0055 0.0222 0.0610 0.1221 0.1832 0.2094 0.1832 0.1221 0.0610 0.0222 0.0055 0.0008 0.0000

Table 2. Normalized coefficients obtained from Table 1B, as described in the text. Note that the sum of the decimals in each horizontal row is 1.0000.

4. Application to Nucleosomal DNA

Figure 3 summarizes the size and composition of DNA segments predicted after limited digestion of the (kinked) left-handed toroidal superhelical form in nucleosomal DNA by the pancreatic DNase I enzyme. We will begin by assuming this molecule to have 147 base-pairs, and to be end -labeled at *one* of two 5' phosphate termini with P32 (indicated by the asterisks).

If the pancreatic DNase I begins by cleaving a kink after the first base-paired segment of B-DNA(i.e.,*B10), *or* after the first base-paired segment of A- DNA (i.e., *A11), oligonucleotides having chain lengths of 10 or 11 will appear with relative frequencies [1 1].

If, on the other hand, the pancreatic DNase I begins by cleaving a kink after the second base-paired segment containing B- DNA (i.e., *B10 B10), or after the second base-paired segment containing both B-and A- DNA (i.e., *B10 A11 *or* *A11 B10), or after the second base-paired segment containing A- DNA (i.e., *A11 A11), oligonucleotides having chain lengths of 20, 21 and 22 will appear with relative frequencies [1 2 1].

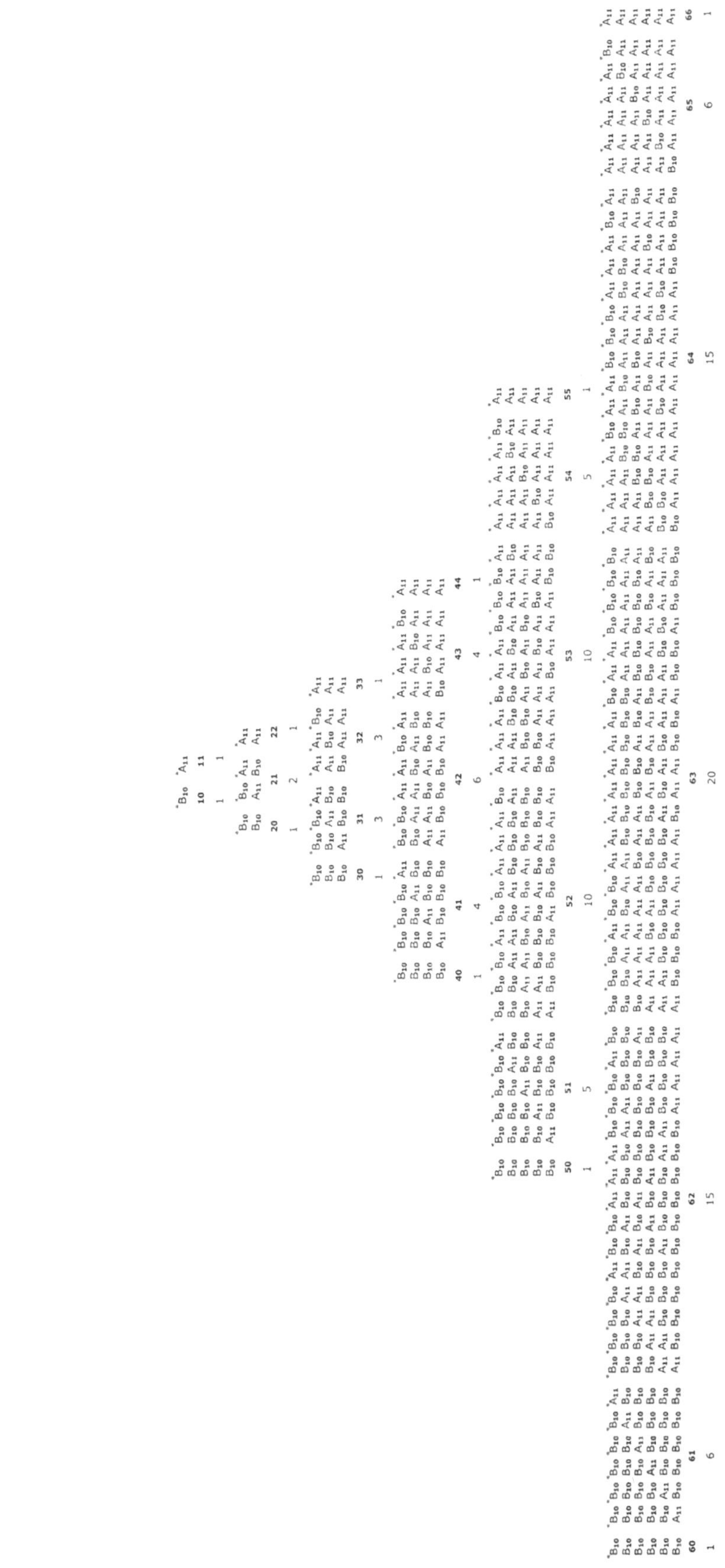

Figure 3: See text for discussion

8

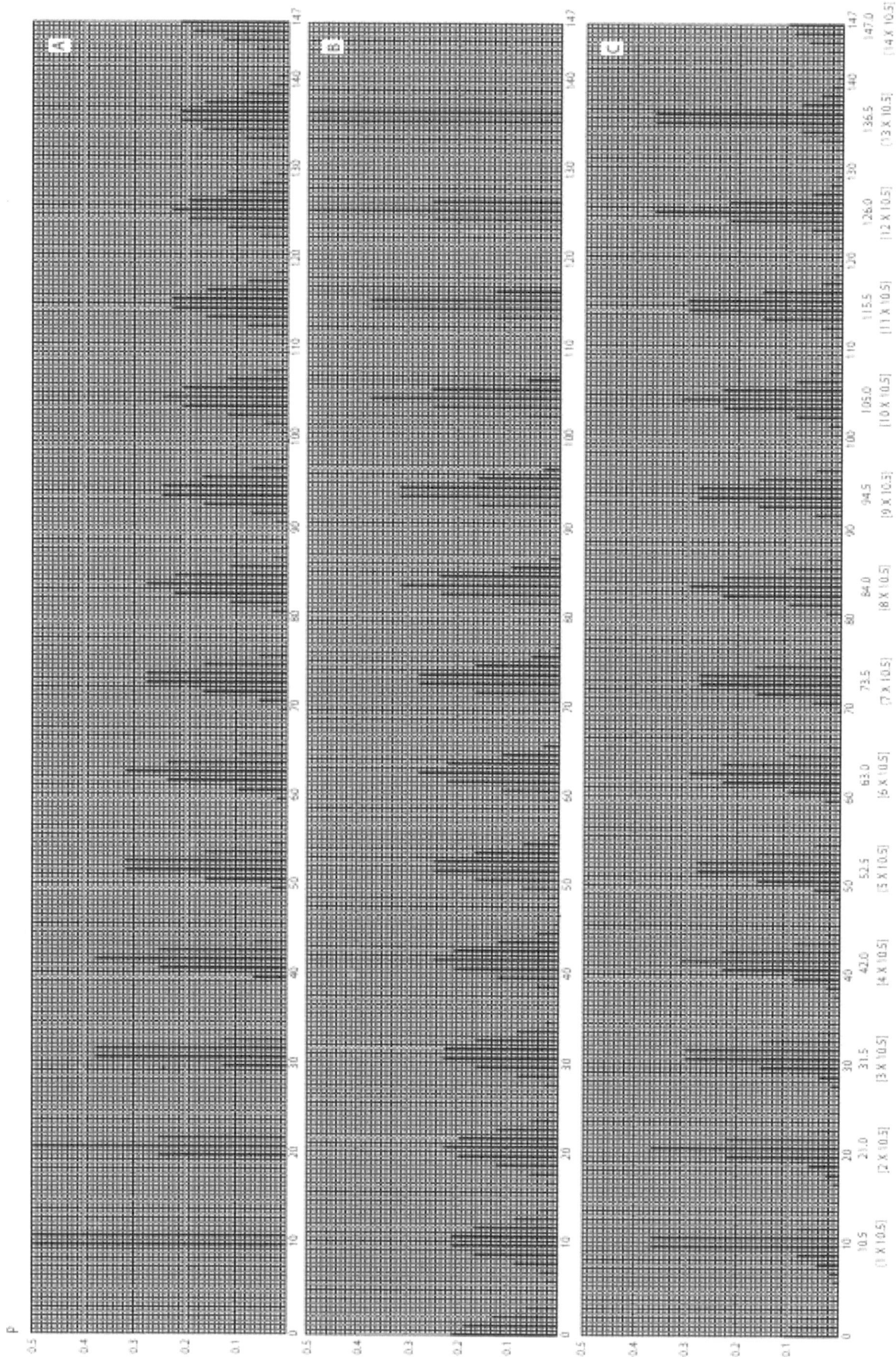

Figure 4: See text for discussion

HELICAL PERIODICITY OF DNA ON AND OFF THE NUCLEOSOME AS PROBED BY NUCLEASES

A. Klug, L.C. Lutter and D. Rhodes, Cold Spring Harbor Symposium of Quantitative Biology 47, 285-292 (1983)

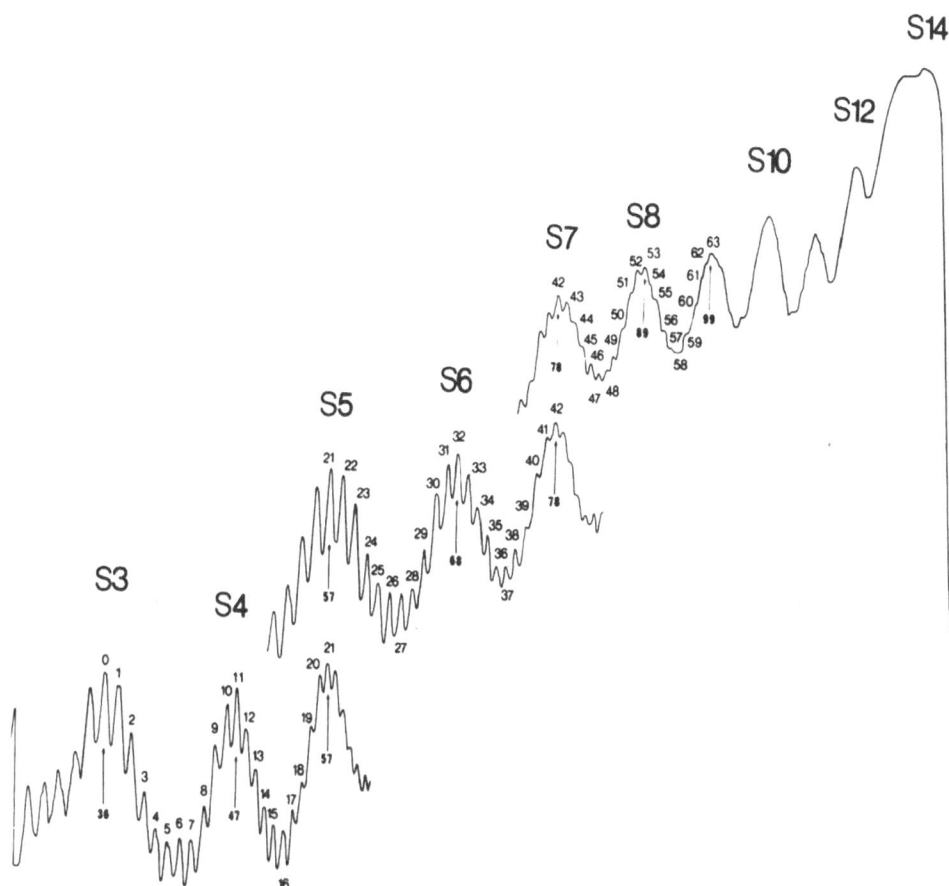

Figure 5: This densitometer tracing shows an electrophoretic pattern of oligonucleotides arising after pancreatic DNase I digestion of a collection of DNA molecules with random sequences, each containing 147 base-pairs and labelled at their 5' ends with radioactive phosphorous -- in the presence of calcium phosphate crystals. The tracing shows a series of maxima spaced 10.5 base-pairs apart; these, in addition, containing finer peaks that differ by one nucleotide. Similar patterns have been obtained with nucleosomal DNA when it is complexed to histone-octamer cores in nucleosomes.

Rather than bending uniformly along its length, the author proposes these observations to reflect the presence of a left-handed superhelical structure being composed of multiple segments, each containing 10 base-pairs of B-DNA or 11 base-pairs of A-DNA, these being held together by 'mixed-puckered kinks'. In both cases, probability considerations predict cutting patterns to be symmetrically distributed around integral multiples of 10.5 base-pairs along DNA, the relative magnitude of the surrounding peaks in these patterns being governed by the binomial distribution.

The fit is not perfect, however, since these experimental results reveal the existence of additional fine peaks between maxima differing by one nucleotide. These can be explained as arising from a limited number of left-handed toroidal superhelical structures that begin with shorter (or longer) B- and A-DNA end-segments. This would have the effect of shifting the patterns (shown in Figure 4A and B) to the left and to the right, causing each maximum in Figure 4C to broaden. This end-effect can explain the presence of additional finer peaks between maxima which differ by one nucleotide.

Finally, if he pancreatic DNase begins by cleaving a kink appearing after the third base- paired segment of B- DNA (i.e., *B10 B10 B10), or after the third base-paired segment containing both B- and A-DNA (i.e., *B10 B10 A11,*B10 A11 B10, *A11 B10 B10), or after the third base-paired segment containing a different combination and permutation of both B- and A- DNA (i.e., *A11 A11 B10,*A11 B10 A11 *B10 A11 A11), or after the third base-paired segment containing A- DNA (i.e., *A11 A11 A11), oligonucleotides having chain lengths of 30, 31, 32 and 33 will appear with relative frequencies [1 3 3 1].

Likewise, it can be readily verified that oligonucleotides having chain lengths of 40 through 44 will appear with relative frequencies [1 4 6 4 1] ; chain lengths of 50 through 55 will appear with relative frequencies [1 5 10 5 1] ; and chain lengths of 60 through will appear with relative frequencies [1 6 15 20 15 6 1].

And so on.

From this it is clear that the size and composition of the nucleosomal DNA segments generated by pancreatic DNase I cleavage in this example obey the binomial distribution. We shall now show that the normalized values of the binomial coefficients define the probability that kinks appear at specific locations determined by n, these kinks being recognized and subsequently cleaved by the pancreatic DNase I enzyme.

Figure 4 summarizes these predictions, both ends of the 147 base-paired fragments having been end-labeled with P32 in this example. The ordinate shows the probability (i.e., the normalized values of the binomial coefficents) that kinks appear at specific locations (determined by n) and then cleaved by the nuclease, while the abscissa shows its chain length position. Since the pancreatic DNase is expected to cause double-stranded breaks each time it recognizes a kink, two different P32 labeled fragments will be generated from a single nucleolytic cutting event. These can be detected by gel electrophoresis as two separate single-stranded fragments, one arising from the 'Watson side', the other arising from the 'Crick side'. For simplicity, we have separated these to show them both individually (Figures 4A and B), and in combination (Figure 4C), the latter graph hav-ing been rescaled for comparison.

Figure 5 shows a densitometer tracing of an electrophoretic pattern of oligonucleotides arising after pancreatic DNase I digestion of a collection of DNA molecules with random sequences, each containing 147 base-pairs and labelled at their 5' ends with radioactive phosphorous in the presence of calcium phosphate crystals (10). The tracing shows a series of maxima spaced 10.5 base -pairs apart -- these, in addition, containing finer peaks that differ by one nucleotide. Similar patterns have been reported with nucleosomal DNA when it is complexed to histone-octamer cores of nucleosomes (6-9).

The data in Figure 5 can readily be compared with that predicted in Figure 4C. In both cases, cutting patterns are symmetrically distributed around integral multiples of 10.5 base- pairs along DNA, the relative magnitudes of the surrounding peaks in these patterns being explained by the binomial distribution. The fit is not perfect, however, since the experimental results reveal the existence of additional finer peaks between maxima differing by one nucleotide.

These can be explained as arising from a small number of left-handed toroidal superhelical structures having shorter (or longer) B- or A- DNA end-segments. This would have the effect of shifting the patterns (in Figures 4A and B) to the left and to the right, causing each maxima in Figure 4C to broaden. This end-effect can explain the presence of the additional finer peaks between maxima differing by one nucleotide just mentioned.

A MODEL TO UNDERSTAND THE HIGHER-ORDER SOLENOIDAL FOLDING OF DNA IN CHROMATIN

H.M. Sobell, B. S. Reddy, K.K. Bhandary, S.C. Jain, T.D. Sakore and T.P. Seshadri,
Cold Spring Harb. Symp. Quant. Biol. 42, 87-102 (1977)

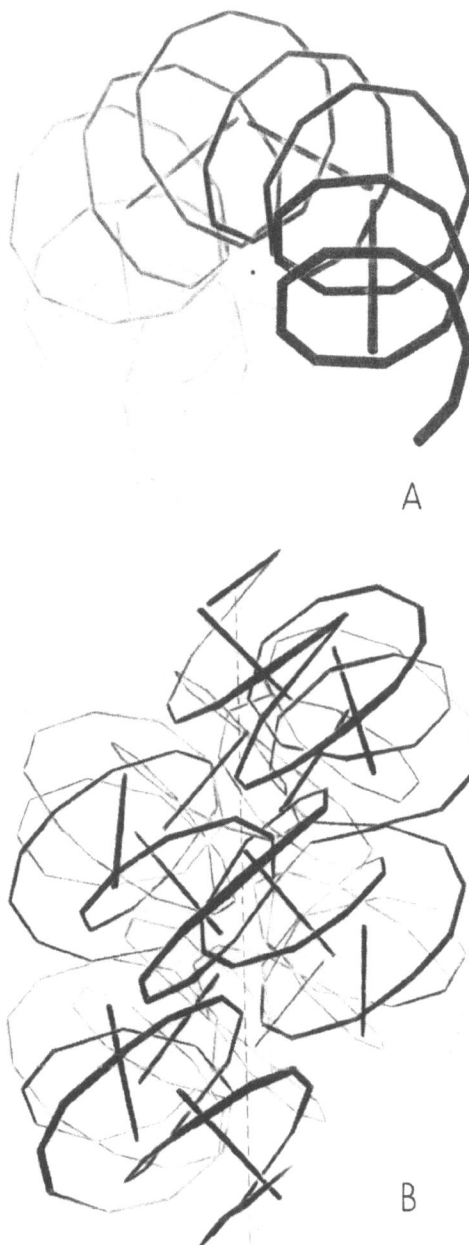

Figure 6: A model to understand the higher-order solenoidal folding of DNA in chromatin. See text for discussion.

12

5. A Model to Understand the Higher-Order Solenoidal Folding of DNA in Chromatin

The presence of mixed-puckered kinks connecting B10 with A11 DNA segments within the nucleosome suggests the likelihood that this same basic structural motif continues to be used when forming the 100 Angstrom fiber. Based on their detailed electron microscopic studies of chromatin, Finch and Klug (1976) and Worcel (1977) have proposed the 100 Angstrom fiber to be transformed (i.e., in the presence of the H-1 histone) into a more compact left-handed solenoidal superhelix, this having a pitch of 110 Angstroms and a diameter of about 300 Angstroms (13, 14). We have explored the possibility that such a solenoid arises due to the flexibility present in the mixed-puckered kink to fold internucleosomal DNA into this form.

Figure 6 shows the results of these calculations. The model shown is a 5.8 fold solenoid, obtained by varying both theta and alpha in the central three kinks within the 60 base-pair connecting region (akin to that shown in Figure 1, panels C and D). The connecting region remains on the outside of the solenoid, with two-fold symmetry being maintained. Structures in this class can readily form the right pitch (110 Angstroms) and the right diameter (300 Angstroms) to fit the electron microscopic data.

An important prediction which this model makes concerns the ability of intercalative drugs and dyes to bind tightly within these connecting regions.

Lawrence and Daune (1976) have described the existence of a limited number of tight binding sites for ethidium in native chromatin, these having binding constants two orders of magnitude greater than for naked DNA. These sites disappear when H1 histones are removed, suggesting a correlation with the solenoidal structure described here (15). In addition, Shen and Hearst (1977) describe psoralin cross-linking with high probability every 200 base -pairs in D. melanogaster nuclear-DNA (16). Our model would predict these sites to occur with highest frequency at integral multiples of 10.5 base pairs (again, these surrounding locations obeying the binomial distribution), and to be highly clustered within the 60 base-paired connecting region. Finally, Cartwright and Elgin (1982) have shown that both the intercalator, 1, 10-phenanthroline-copper (I), and the micrococcal-nuclease cut DNA every 200 base-pairs to liberate the basic subunit structure of chromatin, the nucleosome (17, 18). These data in itself are strongly suggestive that higher-energy mixed-puckered kinks are present within the internucleosomal connecting regions, enabling nucleosomes to form its higher-order solenoidal structure.

6. Experimental Predictions

The model makes a number of testable predictions:

1) The model predicts the number of base-pairs within any given nucleosome to be variable (i.e., lying between 140 and 154 base-pairs; however, having the highest probability that it contains 147 base-pairs), the magnitudes of he surrounding peaks being governed by the binomial distribution.

2) The model predicts irehdiamine and dipyrandium (but not ethidium) -- two steroidal diamines that bind by partial intercalation to the lower energy form of the kink in nucleosomal DNA (11) -- to be competitive inhibitors of the pancreatic DNase I enzyme. Their presence is predicted to suppress the appearance of the cutting patterns observed in Figure 5.

3)The model further predicts ethidium (but not irehdiamine or dipyrandium) to competitively inhibit the micrococcal-nuclease and the chemical-nuclease, 1, 10 phenanthroline-copper (I), from cleaving hypersensitive-sites that exist between nucleosomes in whole chromatin. These regions are proposed to contain higher-energy mixed-puckered kinks in their internucleosomal regions when forming the higher-order solenoidal structure.

7. References

1.   Crick, F. H. C. and A. Klug, Nature 255, 530-533 (1975).

2.   Sobell, H.M., C.C- Tsai, S.G. Gilbert, S.C. Jain and T.D. Sakore, Proc Natl Acad Sci USA 73, 3068-3072 (1976).

3.   Sobell, H.M., B. S. Reddy, K.K. Bhandary, S.C. Jain, T.D. Sakore and T.P. Seshadri, Cold Spring Harbor Symp Quant Biol 42, 87-102 (1977).

4.   Sobell, H.M., C.-C. Tsai, S.C. Jain and S.G. Gilbert, J. Mol Biol 114, 333-365 (1977).

5.   Sobell, H.M., C.C.-Tsai, S.C. Jain and T.D. Sakore, Phil Trans R Soc Lond B 283, 295-298 (1978).

6.   Lutter, L.C., Nucleic Acids Res 6, 41-56 (1979).

7.   Prunell, A., R.D. Kornberg, L. Lutter, A. Klug, M. Levitt and F. H. C. Crick, Science 204, 855-858 (1979).

8.   Lutter, L.C., Nucleic Acids Res 9, 4251-4266 (1981).

9.   Klug, A. and L. Lutter, Nucleic Acids Res. 9, 4267-4283 (1981).

10.   Klug, A., L. Lutter and D. Rhodes, Cold Spring Harbor Symp Quant Biol 47, 285-292 (1982).

11.   Sobell, H.M., PREMELTONS IN DNA, A Unifying Polymer Physics Concept to Understand DNA Physical Chemistry and Molecular Biology, Explanatory Publications, Lake Luzerne, New York (2009)

12.   Arnott, S., R. Chandrasekaran, I.H. Hall, L. C. Puigjaner, J.K. Walker and M. Wang, Cold Spring Harbor Symp Quant Biol 47, 53-65 (1982).

13.   Finch, J.T and A. Klug, Proc Natl Acad Sci USA 73, 1897-1901 (1976).

14.   Worcel, A., Cold Spring Harbor Symp Quant Biol 42, 313-324 (1977).

15.   Lawrence, J.J. and M. Daune, Biochemistry 15, 3301-3307 (1976).

16.   Shen, C.-K.J. and J. E. Hearst, Cold Spring Harbor Symp Quant Biol. 42, 179-189 (1977).

17.   Cartwright, I.L. and S.C.R. Elgin, Nucleic Acids Res 10, 5835-5852 (1982).

18.   Elgin, S. C.R., I.L. Cartwright, G. Fleischmann, K. Lowenhaupt and M.A. Keene, Cold Spring Harbor Symp Quant Biol 47, 529-538 (1982).

www.ingramcontent.com/pod-product-compliance
Lightning Source LLC
Chambersburg PA
CBHW050913210326
41597CB00002B/107